WHEN I LISTENED TO A FARMER

Photographs and Lyrical Stories of
America's Original Entrepreneurs

Written and Photographed by

PETE CURRAN

Hearth & Home Press
MECHANICSBURG, PENNSYLVANIA

an imprint of Sunbury Press, Inc.
Mechanicsburg, PA USA

Ideas, procedures, and suggestions contained in this book are not intended as a substitute for consulting with your physician. All matters regarding your health require medical supervision. Neither the author nor the publisher shall be liable or responsible for any loss, injury, or damage allegedly arising from any information or suggestion in this book.

FIRST HEARTH AND HOME PRESS EDITION: June 2021

Set in Bookman Old Style | Photographed by Pete Curran | Interior design by Chris Fenwick | Cover photo by Pete Curran | Edited by Chris Fenwick.

Cover photo: Nick Raaum

Publisher's Cataloging-in-Publication Data
Names: Curran, Pete, author.
Title: When I Listened to a Farmer / Pete Curran.
Description: Revised trade paperback edition. | Mechanicsburg, Pennsylvania : Hearth and Home Press, 2021.
Summary: Photographs and Lyrical Stories of America's Original Entrepreneurs.
Identifiers: ISBN 978-1-62006-574-7 (softcover).
Subjects: BISAC: PHOTOGRAPHY / Photoessays & Documentaries | POETRY / Subjects & Themes / Family | BIOGRAPHY & AUTOBIOGRAPHY / Cultural, Ethnic & Regional / General

Continue the Enlightenment!

DEDICATION:

Ronald Steinhorst. "Steiny," New London Senior High School Forensics Coach. He helped us find out who we were and what we were capable of.

TABLE OF CONTENTS

INTRODUCTION

Mud. Motor Oil. Manure.

Markers on a farmer's hands.

Revealing the lifelines of a farmer.
Essential GPS coordinates of life.

The city slicker met mud, motor oil, and manure.

On a farm.
On the hands of a dirt keeper.

Ah, farmers . . .
much more than dirt keepers . . . much more than shit kickers.
More than folks who fiddle in the dirt.

The rural fiddler's hand.
Who can communicate silently.
Extended as a professional means of introduction.

A societal connector, threatened today.

The handshake closes the gap between you and me.

Defines who you are.

Who I am.

It gives you 'my word.'

It creates life-long contracts.

An extended hand.

The farmer glances.

A look to his hands.

A swipe to his gut.

Sometimes an aggressive wipe down to his pant leg.

Then, hesitantly, concerned for cleanliness.
A hand still extended.

As a show of respect.

Finally.

Grasped.

I meet the mechanic. Financier. Milker. GPS coordinator. Shit hauler . . . and

the shit-upon.

The farmer.

With many markers over a lifetime.

On his hands.

Markers.

Ours to use.

Those markers. He keeps. Almost permanently stained.

Like the contents in a child's cardboard shoebox of treasures.

A collection.

Of intuition. Tradition. Commitment.

To oneself.

To a mother. A father.

And generations fore, who claimed this dirt.

Who toiled to find a better life.

First, to survive.

Then to nourish.

A Family.

A life.

Is there anyone more noble than the man or woman who is self-sustaining?

Markers. Old. Unchanged.

Perhaps for a reason.

A commitment to his kin and his legacy.

Mud.

The essential element of all farming existence.

There is no survival without the farmer's respect and reverence of the soil.

As a steward of land.

That existence expects success.

Yet can't. If the farmer doesn't steward.

Practices.

For preservation.

Motor oil.

A lubricant to power the agriculture industry.

A reminder to us.

Life keeps moving forward.

Valves. Pistons. And power trains.

To ensure we move forward.

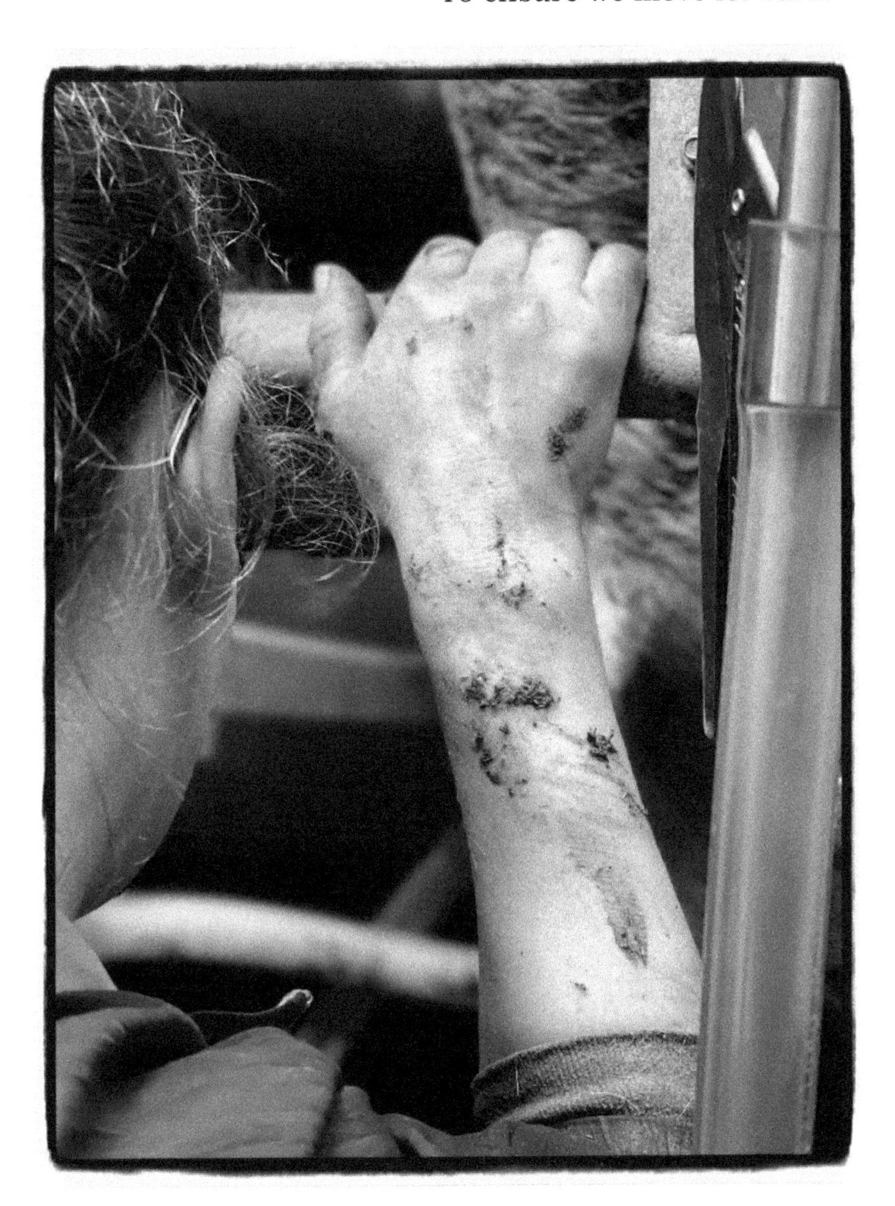

Manure.

No political correctness found here, it's 'cow shit.'

A rite of rural America.

Manure. Rich natural fertilizer celebrated by farmers.

Yet, jeered by those who don't know 'the smell of money.'

Restoring nutrients to soil.

To steward healthy growth and abundant rewards at harvest.

We too, must enrich one's soul.

One's vision.

One's offspring.

With values.

For reasoned evaluation of one's life.

There is a 'way' of a farmer.

Three markers.

Dirty.

Smelly.

Lingering.

Trust these markers we've always known.

Lifelines. Of laborers. Of lives. Well-lived.

A farmer.

Teacher. Tribesman.
Tradesman.

Working.

More like 'living.'

From valley, low.

To ridge, high.

We see the markers.

Left to protect.

The land.

A family.

Solidly. Sufficient!

For hundreds of years so far
. . . perhaps for many more.

A nurturer.

With nuisance.

A farmer.

Welcomes many up his drive.

To his farm.

Where businesses and families grow.

At the ready. With a poker face.

Sighs, then extends the hand to reveal himself.

Connecting.

Sharing.

The essentials.

Of a life.

The mud. Motor oil. Manure.

Ours. To. Embrace.

Can we believe in the simplicity

of grounding our lives?

With ingredients in a simple recipe.

Markers.

Values.

Firmly accessible.

To share the lifelines. That guide us. That comfort us. That sustain us.

Within our reach.

Not new or improved.

Yet, confirmed.

And certified by farmers.

Still patiently tending to the markers.

That define a values system.

That can keep us moving forward.

Through hiccups or hysteria.

Treasured.

STORY 1:

Together is a Timeless Best Practice

Stay home!

March 2020, COVID-19.

America's family farmers were already home.

Mrs. Lins. Mr. Lins.

Son, Steven Lins. Grand-daughter, Jessi.

Three farms.

One family.

Together.

Days grew into months.

Food supply chains were disrupted.

Farmers as 'essential workers.'

Mrs. Lins.

Petite lady.

BIG 'Lins' smile!
(Ditto: Steven and Jessi!)

Rubber boots.

Stocking cap.

Together:

Bike rides.

Boat trips.

Bridges . . . by the burn barrel.

A blueprint.

First impression:
'soft spoken, kind and gentle farm wife.'

And, tough.

Yes, soft-spoken matriarch.

Plus, 'the knife-wielding, chicken-butchering farmer who feeds her family!'

Global chaos. Fear.

Hit the mute button!

Families. Couples. Learning how to spend their time together.

Living.

Working.

Playing.

Going to school.

Together.

At home.

Didn't global chaos send us home in the first place?

Mrs. Lins exclaimed: "We've been practicing for the pandemic for years."

On a farm.

A workplace where families work and live on the same site.

Perhaps the ultimate 'control panel' in a Sociologist Study of 'work/life balance.'

Where a kitchen delivers family meals and business deals.

Where a barn houses income sources and places for kids to play.

Simple farmhouses, just feet from a barn. A farm. A business.

Where tractors, milk trucks, livestock trailers, and manure spreaders pass daily.

Constant movement and motion, part of the daily chaos of agriculture.

Can chaos bring opportunity?

A farm.

A blueprint to thrive.

The recipe for life is seasoned by nature, faith, and tenacity.

Mrs. Lins:

"We took time for each other."

Steven Lins:

"I got to eat lunch with my kids every day."

Jessi Lins:

"Nathan and I want to raise our family the same way we grew up . . ."

Bike rides.

Five kids.

Mom and dad.

Up and down country roads, around the 'block.'

Together.

Boat trips.

A family of seven.

In a boat.

On a Sunday.

For fun on the Crawfish River.

Together.

Bridges . . . by the burn barrel.

Before more uncertainty finds us, how are you spending your time?

A mom of five.

Randy. Lost.

In a farming accident.

Together in spirit.

She built a bridge.

Over a creek. By the burn barrel.

To be alone.

And together.

To connect.

Comfort the mind.

Refresh one's engine for living.

A break in the day.

Bike rides.

Boat trips.

Burn barrel bridges.

Old fashioned monikers of a time gone by or timeless cliff notes for a healthy family?

Back at it!

Work.

Living.

Together.

"Together . . . is ingrained in farming."

Chores. Of course.

Mowing the grass.

Picking weeds in the garden.

"Putting the five of them in my Ramble to get groceries and seven gallons of milk for the coming week."

Together.

Family.

"We'd make hay with the neighbors . . . vacation at home because we had animals . . ."

Simple things.

" . . . go to church . . . host 4-H class . . . a quick break to the creek to hang out on a branch . . . cut beans during canning season . . ."

Pray.

Play.

Persevere.

Together.

Baling hay.

Brothers & sisters.

A rite of passage for farm kids.

A test of fortitude.

For mind and body.

Your childhood.

Not so bad.

It's time for some family bonding.

Use the skills you were taught.

Life.

Rough Road Ahead.

Reality.

Our responsibility.

Real world insight.

'Cause there ain't no sanitized world.'

Together.

"Always someone to help."

Farming.

A best practice for times of chaos?

Tell'em.

Yep!

Tell'em.

Put down the phone.

Take off the headphones.

Stash the game controller.

2020 . . . a gift?

Shut down the uncertainty.

"All hands on deck!"

STORY 2:

Weathered.

Listen to the slow creaking of the screen door, opening on a muggy summer day.

Weathered, it dutifully hangs on one hinge.

The farmer's arm lay on the bottom of the screen.

Pushing.

Open.

To get.

To work.

A few yards to the barn.

Past the fieldstone-lined well.

From his home.

WHACK!

The screen door slams and sounds the start of the day.

Farmer. Farm wife . . . there they go!

Work to be done . . .

Some 55-plus years later.

The home, farm buildings, and door are discovered.

When a young farmer tells me, "I need to show you something . . ."

Hanging by one screw of a metal hinge.

Many storms.

Still hanging on.

The empty farmhouse.

Its screen door hanging on an angle.

Complete with rusty c-handle.

And eyelet hook to secure it.

Lays open.

To reveal.

The markers to a well-lived life.

Like a farmer.

Who's home and business.

A farm.

Are on display for all who pass.

To view.

To comment.

To wonder.

A wooden screen door.

A weather-worn farmer.

Each with roles in life.

One to let in gentle breezes and keep out pests.

One to rise and rest with the sun to steward a land.

Roles tested by nature.

Roles tested by chaos.

Opening.

Closing.

Rising.

Resting.

All necessary actions in the recipe for life.

Where thunder shakes more than the walls of a home.

The screen door and the farmer.

Leaned on at times of pause.

When the farm wife wonders . . . if she can 'do' this.

She did so!

The farm wife and mother of this home.

A commitment to family.

At home until an adult son died in the 1960s.

55 years later.

The screen door.

The farmer.

Solidly.

Hanging on.

To show us:

The simplicity.

Of a well-meaning life.

Valued.

For the family values nurtured.

Knowing why we work.

Responsibility.

Reality.

Valued.

Grateful for life, given through exertion.

Results.

Realized.

Valued.

Perspective. To equip a generation.

Recognized.

Recalled.

WHACK!

Romanticized and recollected.

Farming.

A whack in the head.

Brings meaning over pain.

Encouraged by the wind.

The screen door opens.

The farmer catches his 'second wind.'

Both have work to do.

And, so do we.

The answers to many of life's questions are found in the commitment of the simple, wooden screen door that doesn't let go.

Just like the traditional teachings kept safely in our heart.

At home.

Where its roles were many.

Like ours.

So seemingly simple, and often overlooked for their contributions.

The screen door allows us to see the comings and goings of a life.

And to open and close, as desired.

To stay true.

After a storm.

We can open our back door.

And let the screen door share the cooling breeze from the rain.

The headwinds of the 'un-experienced' with -- new narratives -- suggest we change.

No need.

We can calm the discomfort.

By staying true to the tailwinds of family values that have delivered so consistently . . . for so many.

Our lives may wear and tear over time; those who are weathered sustain value.

STORY 3:

Fire. Death. Birth.

Events.

Tear us apart.

And . . .

Bring us together.

On a farm . . . and in life . . .

Fire.

Ravaging.

Warming.

Fierce.

Calming.

Callous.

Comforting.

Death.

Disconnecting.

Defining.

Silent.

Celebratory.

Heartbreaking.

Healing.

Birth.

Messy.

Miraculous.

Reminiscent.

Renewing.

Terrifying.

Teaching.

Events to . . .

Remember.

Forget.

Big.

Small.

Doubt.

Determination.

Tools that help us.

Backhoe.

Shovels.

Hands in gloves.

Family.

Friends.

Neighbors.

Embrace reality.

Fire in the goat shed.

A farmer saves.

Then nature takes.

Loss.

Love.

A community gathers.

All hands on deck.

Support engulfs the chaos.

Charred wood.

Metal.

Plastic.

Garbage.

A tricycle.

Horseshoes.

Tobacco harvesting axes.

Pails filled with nails.

Things (full of memories.)

Picking up.

Putting away.

Smokey smell.

Sooty faces.

Closure.

Smiles.

30+ folks clear the rubble.

Concrete floor revealed.

A vision of a future 'shed-raising.'

Release.

DEMONS
Deerfield
TRACK & FIELD

Wonder what the farmer thinks?

Folks directed to the barn.

Goats in labor.

Life . . . goes on.

Renewal.

Despite the fire . . . it's still birthing season.

Twins.

Seven new babies.

Farmer nurturing life.

Warming light.

Babies in a pen.

Feeding one on his lap.

Smiling.

Mentions of 'what could have been worse.'

Grateful for the help.

Resilience.

Learned.

Necessary on a farm.

Necessary for life.

STORY 4:

A Door to a Circle of Life.

Fresh, clean snow.

A white palette on the earth.

The red, worn door.

Noticed.

For the first time.

Not my first visit.

A simple, four-panel door.

An entrance and exit to an Outpost of Industry.

A farm.

A door to a barn.

A door to work.

Morning. Noon. And night.

Often in the dark, the doorknob is turned another notch . . .

Silence.

Standing still for a moment.

I saw a farmer's circle of life.

Through the people who opened that door in its 103 years of service.

To this barn.

To the farmer.

None of us alone.

I saw a circle of life.

Co-workers. The farm dog. Family. Friends.

Those who cobbled fieldstone and mortar.

Into the foundation of a building.

A barn.

An outpost of industry.

A home.

The foundation to a life.

Where neighbors worked together at harvest . . .

Grandfathers, slower yet still moving with wisdom . . .
Up on the tractor every day . . .
Grandmothers . . .
perhaps on a final visit to check out the barn, to say goodbye to a life so loved
with a touch to the fieldstone foundation on the palm of her hand.

A young granddaughter, standing in the gutter of the stanchion barn watching
grandpa milk cows . . .
energized kids causing havoc . . . laughing . . . getting yelled at . . .
doing chores at an early age.

The vet,
Tends to the sick and encourages the new.

The milk hauler,
Collects the works of the family's labor.

A one simple photo of a door opened folks' recollection of their own circle of life.

The cycle of life reaches the end . . . a soul who created a career to steward the land . . . that built a business . . . that created a family . . . that taught the lessons of a life . . . to sustain one's self.

The farm hand, hired to help . . . or replace a child who ventured off . . . to their own farm . . . and their new family.

The calf feeder, often the wife, a nurturer.

The hoof trimmer.

The kids learning how to handle a heifer to show at the County Fair . . .
in hopes of a blue ribbon . . .

Field stone foundations.
Weathered red barn boards.
Structures of an industry
Much more than buildings.

The door to a barn where, before tending to the animals,
a playful newlywed couple, the day of their wedding . . .
A granddaughter swings on a pole,
falling into her first kiss as farmer and farm wife

A circle of life, where

Practices are modernized.

Doors are replaced.

Foundations remain.

A door to a barn.

The entrance to values of work ethic and dedication to family.

Enduring.

Years of reviews

From 3rd and 4th generations

Young farmers drawn to this door.

The farm.

Where a circle of life is shared

with each turn of the doorknob.

Repeated.

Unless, unnoticed.

STORY 5:

Trespassing on a Thursday.

Drawn, like a beacon to the shore.

A large farmhouse.

Perched atop a hill at the end of a long, gradual climb up the driveway.

Etched in time.

Now ‘nestled’ next to an Interstate Highway

in Wisconsin.

I’ve flown by this house for years going 70 mph.

Always glancing up at the white farmhouse.

Wondered often who lived there.

Windows broken or missing.

Just dark, black spaces to match the roof.

Like a deep, dark hole.

Empty.

Still drawn to this house.

Trespassing.

A bit of fear.

I walked around the bolted gate.

Approaching.

The home grew bigger.

A beaten down path through weeds.

Veering right.

Destination.

Unknown.

To see a farmer's life experience.

The farmhouse.

A grand centerpiece of this family farm.

A vast 'search engine'

Trespassing.

Another black hole.

A root cellar?

The door laying on the ground.

Peering down the cement stairs.

Pink walls. White & black graffiti.

I dare not go down.

Too many scary-teenager-horror-films . . .

Shattered windows and wooden casings litter a timeless foundation.

Up the stairs.

A wide-covered porch.

More graffiti.

Eerily colorful.

Those shattered windows ice my steps.

Vandalism. Endless. Pointless.

Frozen hearts. Mine broken.

Knowingly drawn to trespass today.

Broken heart torn but not frayed enough

to still gather.

Insight.

From tarnished framework.

Standing still.

Listening.

Looking.

Seeing.

. . . the family . . . hearing chatter . . . erasing the ugly tarnished reality . . .

To perhaps, 'rekindle the value of one's life.'

By connecting the broken pieces of this home.

To meaning.

To remnants and remembrances.

Of stuff and times, now considered old and 'old fashioned.'

Noticed though, a marker of time . . . cracked plaster ceilings and walls.

A wooden divider between parlors.

Decorative columns.

Turning, the farm wife welcoming Sunday guests.

In an apron

Like grandma . . .

On the covered porch.

Shade and a breeze.

Overlooking a vast rolling hill . . . fertile, flat land down below.

On the stoop . . . kids giggling. Women in motion in the kitchen. Smiling.

Men anticipating harvest, resting in the parlor.

Anticipation.

And, yes!

The vandalism reminds.

Pain. Doubt. Wonder. Prayers. Hope.

Just like us.

Authenticity.

Trespassing.

The hiss of spray paint.

A back kick through the dining room door.

A rock crashing against the picture window.

Broken.

Windows.

History.

Hearts.

Apathy.

Vandalized by a shallow sense of one's worth.

Appreciation.

Trespassed and tagged in one's mind to share worth . . . in structure . . .

A centerpiece of family values?

Or . . . just a house.

On a farm.

Where folks work in the dirt.

For betterment.

In a new world.

Still grounded in tradition.

Discovery.

A home.

A barn.

Only yards apart.

Firmly planted next to each other.

Providing.

Protecting.

Possibilities.

The centerpiece of American Family Values?

Today: Opinions. Rebuttals. Demands.

Our world, 'so digitally connected.'

And seemingly 'extremely disconnected' to ideals that created a country.

A way of life.

A way of common purpose.

Trespassing.

The parlor. Dining room. Kitchen.

Broken! Spray-painted! Maligned!

Disrespected!

Reject a model for living whose markers have stood the test of time?!

Where else do home and industry share ground?

Where families.

Live.

Work.

Play.

Putting on their 'Sunday Best'

after getting splashed by cow shit in the barn.

Keep working and knowing one's faith also needs tending to.

Our world spinning faster than ever.

Smart phones.

It's time for the guests to head home.

What was the pause taken on this porch?

At sunset.

Trending: 'Farmhouse Style.'

Fascination with the rusty contents of a farm?

More likely.

A desire.

To fill one's own black hole.

Empty. Scary.

Craving.

Comfort.

Calm.

Of grandmas. And grandpas.

Parents.

Who lived a way . . .

Work.

That tires.

De-stresses.

Accomplishes.

Play.

As family.

Together.

Outside.

Around the kitchen table.

Living.

Alongside.

Father. Daughter.

Co-worker.

Mentor.

Through an honest days' work.

Responsibly.

Life has changed.

All things.

Need.

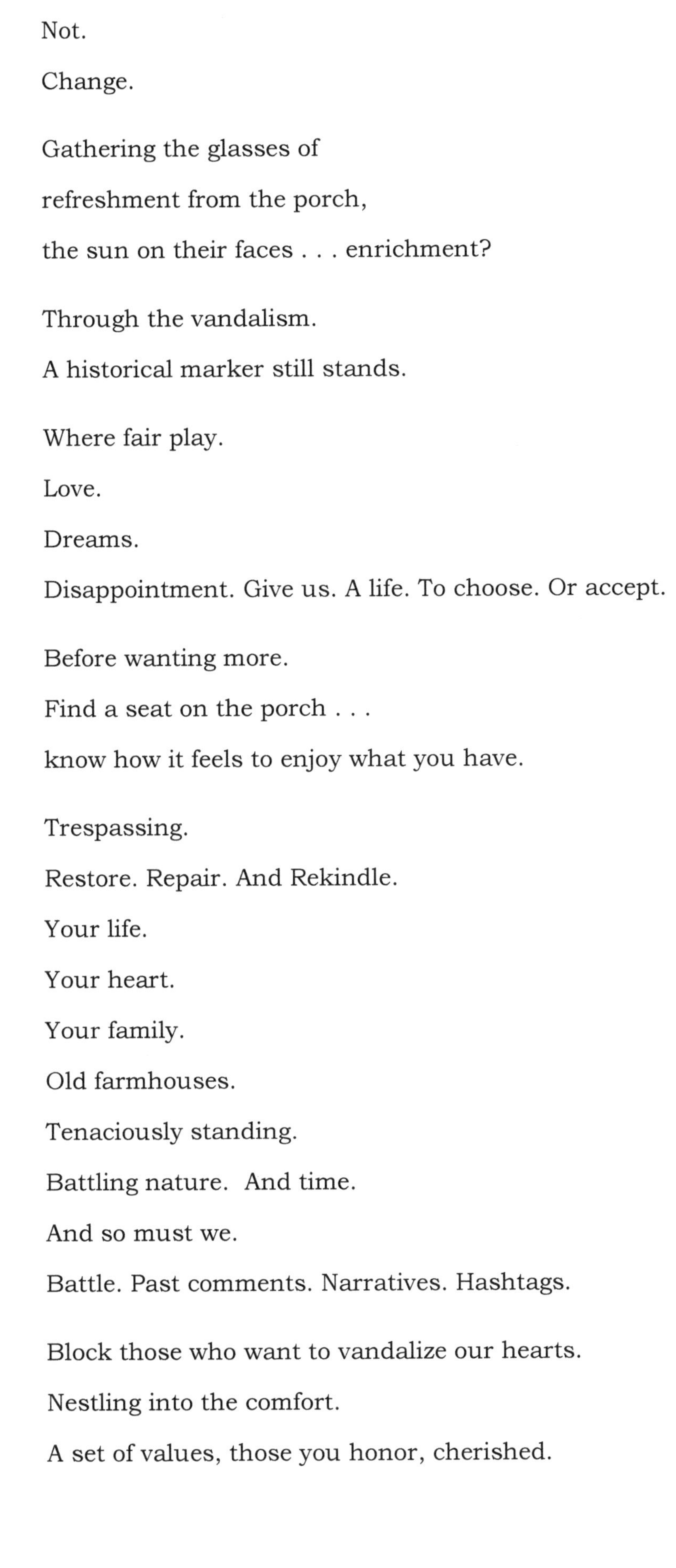

Not.

Change.

Gathering the glasses of

refreshment from the porch,

the sun on their faces . . . enrichment?

Through the vandalism.

A historical marker still stands.

Where fair play.

Love.

Dreams.

Disappointment. Give us. A life. To choose. Or accept.

Before wanting more.

Find a seat on the porch . . .

know how it feels to enjoy what you have.

Trespassing.

Restore. Repair. And Rekindle.

Your life.

Your heart.

Your family.

Old farmhouses.

Tenaciously standing.

Battling nature. And time.

And so must we.

Battle. Past comments. Narratives. Hashtags.

Block those who want to vandalize our hearts.

Nestling into the comfort.

A set of values, those you honor, cherished.

Reset your table

with family values

center of your life.

STORY 6:

Eternity

The landscape of rural Wisconsin is dotted with them.

Steeples.

Silos.

Exaltations!

Heritage. Values. Work ethic.

Expressed by taller-than-life reaches into the sky.

Structures that can lift us up.

'Faith Places.'

Churches and Farms.

Where folks enter to work.

Where families grow.

Radio blaring.

Adrian's dairy barn.

Sales call.

Fatherly figure.

Cordial.

Calming.

Silence.

Adrian's words were interrupted

pauses and the closes of his eyes.

In his farmhouse parlor while sitting together,

he was in pain.

Yet comforted.

Home.

The 'Faith Places.'

Mimicking each other.

Outside: Distinct architectural features.

Steeped in tradition.

Inside: Vast. Expansions.

That have heard many a prayer.

Where one can make choices.

Or not.

Create a plan.

Change course.

Stay true.

Where life is celebrated.

And loss is learned.

Where we make deals with God.

Or ask for help.

Centered.

Pews and stalls.

With an aisle.

To walk things out.

And lofts.

One for choirs to sing praise.

To uplift. A glorious noise.

One, a hay mow.

To note bounty. To confirm preparedness.

Often. Both quiet.

Spaces.

To amplify silence.

To accentuate life.

Both. With a capacity.

Of storage.

Of thought.

Of hope.

Singing, mooing, and diesel engines running . . .

white noise to distract the mind.

Two journeys.

Adrian knew.

Some still finding out.

Only took a moment to see Adrian's peacefulness.

Leaning on his white pickup.

Me on the tailgate.

Listening.

We figured out many of life's woes.

Always.

Feistily-patient.

Stopped to check in on him from time-to-time.

Weaker.

Yet smiling.

Invites me to sit down.

He stopped treatment.

Morphine.

Pain.

Pause. Eyes closed.

In the parlor.

Pants dirty.

Sitting on the floor.

Looking up like a child.

In awe.

Of wisdom.

Of patience.

Of fate.

Talked about the market.

A story from the business page.

Asks me where I think the grain market's going . . .

It was clear then.

Where he was going.

Adrian lying back on his recliner.

Peers out the window.

Asks me to go give Jon a hand.

"Come up to the loft."

Never been up there.

Climbing the ladder.

Overwhelmed.

The vast cathedral of the barn.

Illumination.

Aura.

A vision of Adrian's destination.

Walking exhilarated and exasperated toward Jon.

Standing on an altar of hay bales.

Reaching up.

Connecting the chain of a conveyor.

To bring newly-harvested bales up into the loft.

Returning to the parlor.

Time to leave.

"Are you saying goodbye?"

Adrian smiled and said, "Not yet."

Yet the revelation in the loft.

Was no coincidence.

Clarity.

"What words of wisdom do you have for me . . . and others?"

Adrian replied, "I believe there is an eternity."

I shook his hand.

And said, "I'll see you later . . ."

Faith offers serenity.

STORY 7:

Yardstick

Life measured.

The olden days . . . dust storms and . . .

'An honest day's work.'

'Fair play.'

'Giving thanks to God.'

Today . . . deadly disease and . . .

'Social media Likes.'

'Vacation destinations.'

'Our new stuff.'

Our modern life.

Perhaps a time to critically reflect on what we measure.

Re-boot . . . re-evaluate . . . restore . . .

To increase joy.

And good health.

Our world always spinning.

Sunrise. Sunset.

Your life . . .

Spinning too fast . . .

Just hanging on . . .

Find your yardstick.

Measure twice . . .

Confirm.

Commit.

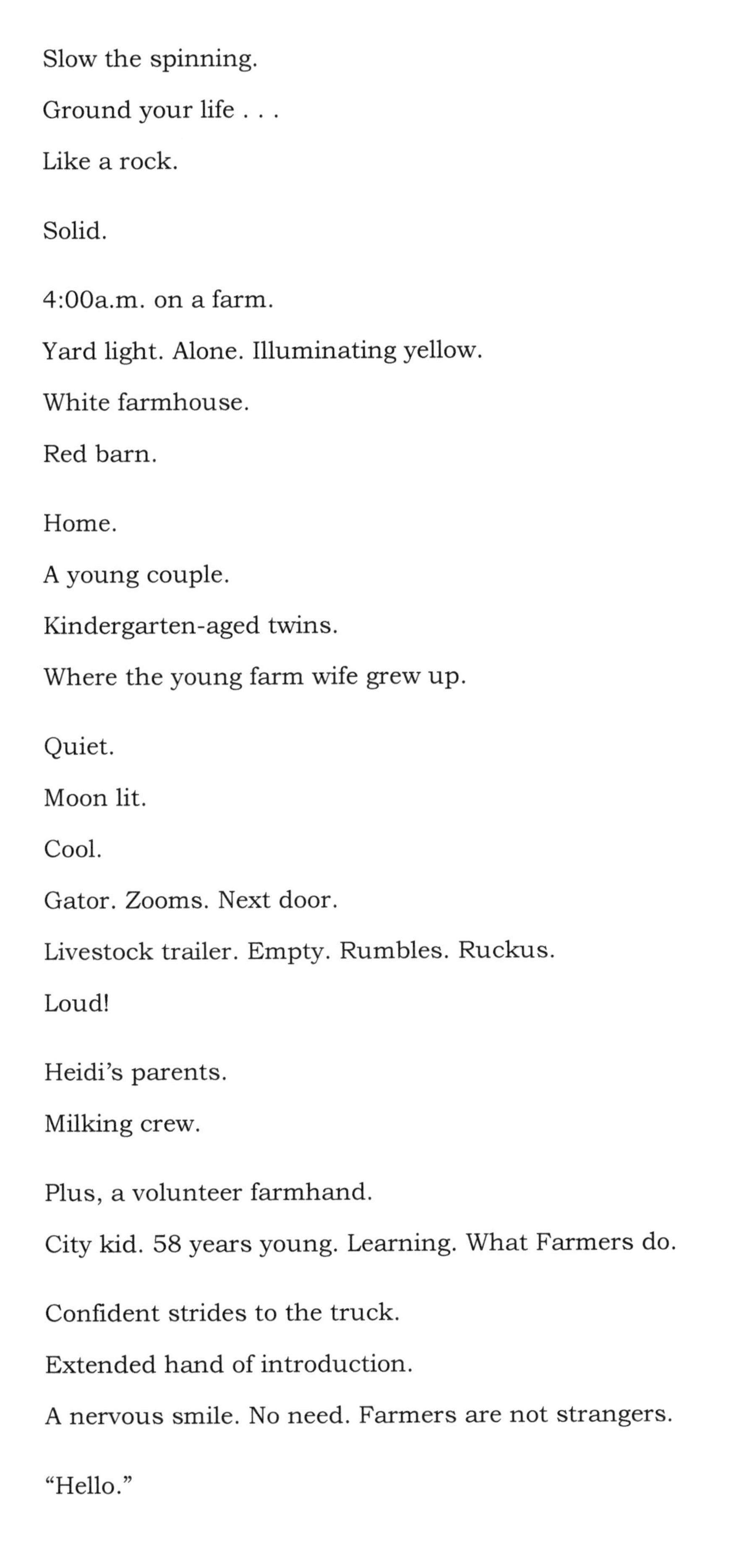

Slow the spinning.

Ground your life . . .

Like a rock.

Solid.

4:00a.m. on a farm.

Yard light. Alone. Illuminating yellow.

White farmhouse.

Red barn.

Home.

A young couple.

Kindergarten-aged twins.

Where the young farm wife grew up.

Quiet.

Moon lit.

Cool.

Gator. Zooms. Next door.

Livestock trailer. Empty. Rumbles. Ruckus.

Loud!

Heidi's parents.

Milking crew.

Plus, a volunteer farmhand.

City kid. 58 years young. Learning. What Farmers do.

Confident strides to the truck.

Extended hand of introduction.

A nervous smile. No need. Farmers are not strangers.

"Hello."

Silence.

Straight face.

Alan.

Pointing to the gator, an introduction to his wife, Fran.

Walking into the parlor.

Shared the 'just-kick-me-out-when-I-get-in-the-way' suggestion with a smile.

Alan.

Still.

Straight face.

Silent.

Grabbing iodine to dip the teats, putting on milkers, and spraying hooves.

To contribute.

Still tension in the parlor. Or is it me?

Started asking Fran questions.

Courage up . . .

Turning to Alan, “What is best and hardest thing of being a farmer?”

No longer silent . . .

“My parents gave me the opportunity to farm. I worked hard and took it.”

Straight face, shattered . . .

Talking faster . . .

Smiling . . .

Spark in his eyes . . .

Alan announced the yardstick used to evaluate his life.

The Four Fs.

Excitedly he recalled them . . . starting . . . stopping . . . starting again to get the order right.

His wife, Fran calms him . . . jumps in.

Reminds him of the first F.

Faith.

"We have no control over anything . . . so we give it to God.

Think about this . . . we have all this . . . just because two people fell in love."

Family.

"Blessed to have our four girls!" exclaimed Alan.

Electricity in his eyes,

so noted.

Fran adds, "and four son-in-laws."

Joy.

The next generation shares the same values that built our family.

Tribal knowledge.

Apprenticeship.

Make joy.

America? Need joy?

It's here.

Alan's been milking on this farm since he was 10-years old. Now Heidi and her family are making a life here.

Joy.

Farm.

Alan and Fran, both farm kids.

Work ethic . . . rock solid.

A farm shows the value of working together . . .

To get work done.

. . . and being together . . .

Fran.

"Franny!"

Alan's wingman for the last 42 years.

His exuberant expression refreshes.

Co-worker.

Friend.

Wife.

The rock of his grounding.

Faith. Family. Farm. Fran.

Measures of one's life.

Alan's yardstick.

Blurted out with certainty.

Uncover your gratitude.

What do you measure?

Ground your life.

STORY 8:

Beating Heart

Blessed?

Or burdened?

The farmer.

The legacy keeper.

See the beating heart . . .

3rd generation.

Nurturing.

4th generation.

See the beating heart . . .

In the homestead where grandma and grandpa lived.

On the road to uncles neighboring farms.

Since 1939 . . . says the sign.

Saved.

Proudly displayed.

Preserved.

To remember.

The first beat.

That created this heart.

This legacy.

Choices. All of us.

Misperceptions.

Ours.

Others.

Accurate?

Or assumed?

Don't judge the steadfast focus of a farmer.

Four years.

Sharing time as baseball parents.

Saw Bruce, less than 10 times . . .

In jeans . . . clean baseball cap.

Watching his sons.

His heart.

From the fence.

Their hearts.

From the warm-up circle.

See the beating heart . . .

Bruce's dad. So focused.

No time for play.

Bruce wanting to shoot hoops with him.

Desires.

Didn't happen.

So, he built a field of dreams.

Little league field.

Backstop.

Barn.

Flag.

Fence.

Outfield.

Pasture.

Together.

Apart. ('someone needs to milk the cows!')

See the beating heart . . .

Satisfying?

Or sacrifice?

Crazy?

Or calming?

Farming.

Flailing . . . failing . . . forgotten . . .

Not. Ours to judge. What we don't know.

Pressure?

Or preservation?

Living.

Constantly working.

Carrying on.

See the beating heart . . . of those. Who know.

Simple?

Or uninformed?

Same spirit.

Same hope.

The beating heart of the new generations.

Farming in his footsteps.

Or not.

Choices.

Buildings.

Ways.

To live.

Symbols.

So profound.

Symbols worth saving.

Respect.

Pride. (especially in family)

Perseverance.

Their value never fades.

See the beating heart . . .

We'll always need.

The old fashioned.

Never changed.

Heart. Of a farmer.

Focused. In. Belief.

STORY 9:

The Original American Entrepreneur

Colleges offer degrees.

In entrepreneurship.

Farm kids . . .

Entrepreneurs:

By age 8!

Generations . . . entrepreneurs.

Gain independence.

Choosing what to do.

Vs.

Being told what to do.

Ingenuity becomes necessary.

In work on a farm.

In life.

Independence with expectation of self-sufficiency.

Key skills:

Quick assessment.

Problem-solving.

Action.

Absolutely!

Margins . . . slim.

'Mondays' happen every day!

Farm equipment is a student of Murphy's Law . . .

Farming's a shit show!

Farming's the ultimate 'do-it-yourself-project.'

PIONEER

HARVESTORS

Like life.

For life.

From scientist to welder to plant pathologist to cow pattie hauler,

a farmer casts a wide net of skill sets.

Farm skills are survival skills.

Mending machines.

Can . . .

Mend minds.

Accomplished!

Life breaks.

React.

Assess.

Act.

Expect it to break again.

STORY 10:

Please Stop Talking

Really.

Please stop talking!

Relax.

Rejuvenate.

One rung at a time.

Higher.

Closer.

Only the second install of "Stars & Stripes" atop a grain bin.

110 feet in the air.

On Steve's farm.

Reality.

Beautiful blue sky.

Light breeze.

Sunny.

Repeated.

"I'm afraid of heights!"

Required action.

Up the steps.

Up the ladder.

Nervous.

Recall fear.

90 rungs ahead.

Recant confidence.

Steve shouts out how my body may react to fear.

As anxiety increases.

Shouting back:

"PLEASE STOP TALKING!"

Stop reciting self-proclaimed limits in your life.

Re-establish confidence.

In yourself.

In your family.

Refocused.

One rung at a time.

Climbing a grain leg.

Climbing life.

Relief.

The top.

The platform.

Reward.

Refresh.

Catch your breath.

Relegate limits to the rubbish heap!

Remember trying . . .

Riding a bike.

Jumping into the pool.

Getting on the Tilt-a-Whirl . . .

Recall.

Failure.

Resilience.

Pain.

Results.

Reality.

Please stop
talking!

Redirect 'can't.'

Re-do.

Repeat.

Relegate.

Family reluctance.

To try.

Remove doubt.

In living.

Re-kindle . . .

A required.

Reverence.

For living.

Recycle.

Reimagine.

Realize.

A life.

Comforted by re-takes . . .

One day at a time.

Please stop talking!

Release.

Confidence.

Recollected at
family reunions.

Grandpas and
Grandmas . . .

No age. No skill.
No lack of
courage . . . .

No remnants of
limits.

Remembered.

Please stop
talking!

Return.

To independence.

Keep climbing.

To replenish and renew.

Relish this life!

Accept your ability to succeed.

The upside of risk . . .

STORY 11:

Moments

Haily.

Born in a hailstorm.

Moments are catastrophic.

Moments are blessings.

Growing up.

Farmer.

Father.

Leads her up the stairs to top of the grain bins.

Perspective of life . . .

Wonders on the landscape.

Sunset colored skies.

Scouting crops for assessment.

Listening to the quieting birds.

Discoveries.

And possibilities.

Atop the grain bin.

Student.

Teacher.

"There's got to be a higher force . . . this is too big!"

Wedding day.

Atop a balcony of a barn.

Haily arm and arm with her father.

She takes the first step.

Leading her father down the aisle.

Steve.

Humbly by her side.

Escorting joy.

Head down.

Tepid smile.

Haily.

Holding up her mentor.

Glowing smile.

A silhouette of confidence.

From moments remembered.

Moments defined.

Lasting.

On the farm.

In life.

"It's not a matter of 'if' you are going to have problems, but when."

Equipment. Down. Again.

Woodchuck stuck. In grain auger.

Drought dries up income.

Grain dryer stops drying.

Job is eliminated.

World closed. COVID-19.

Moments spent teaching.

Her father, Steve says.

When your job is a farmer, there is no time clock or set hours.

We can't walk away from a problem, it's ours until we fix it.

Living and working on the same ground.

We don't get to put away problems until tomorrow . . .

Moments recognized.

I could see that Haily was absorbing all of this.

Awe in her face, body language.

Moments remembered.

I got to see my Dad 'happy.'

He's a hard worker that loves to farm."

"Farming is who we are!"

Haily fiercely exclaimed.

Moments are lasting.

"Your everything goes into that."

Each year is a new start.

Hardship.

Happened.

Circumstances change.

Just like the seasons.

"It could be our best year yet . . . "

Your life.

Problems

A constant.

Calamity?

Or . . .

Celebration?

Moments are reactive.

Perspective

A choice.

Coping skill!

Born in a hailstorm.

Haily's confrontations with chaos comforted by the quiet demonstrations of her father.

She was shown and believes,

"No matter what, we'll figure it out."

For those who chose a career that relies on Mother Nature, farmers know her best.

Moments reveal.

Farmers.

Solving problems.

Daily.

On their own.

Farmers.

Trust them.

Moments are unexpected.

You, too, must solve problems.

Beyond duct tape

Farmers are the ultimate problem-solvers.

We can weather any storm.

No matter what.

Give your family the duct tape and gumption . . .

To tape it up . . . to figure it out.

Terrifying?

Or . . .

Temperate?

Moments. Reusable.

STORY 12:

Cow Pies

Splat!

Splash!

Plop!

Sounds of fresh cow pies straight from bakery

a barn.

"Take cover!"

Take a sample . . .

This ain't no pie judging contest.

It's Fecal Sample Day.

Knee-high green rubber shit kickers

aka . . . muck boots.

To see if what goes in a cow

Is tasty and fulfilling.

By checking what comes out . . .

Armed with an ice cream pail.

We look for fresh cow pies.

Aka, 'shit.'

Splat!

Splash!

Plop!

"There's one . . . "

Quality control . . . of the recipe.

Of feed.

Essential for energy . . .

To produce milk.

Empty bucket.

"Check."

Clean hand.

"Check."

I dug right in.

Documented.

My boss laughed and said, "Why didn't you were gloves?"

I replied, "I've never felt manure and needed to know how it felt."

Even the cows wondered what I was doing.

A checkmark gained on my ‘Ag Experience Bucket List . . . ’

Laboratory Official.

Gather three fresh fecal sample.

Mix.

Yep, by hand.

Place sample in bottle.

Seal.

Identify.

Deliver to lab.

Let the judging begin.

Reality check:

'Feel the work.'

To learn.

To know.

The Work . . .

Dirt.

Shit.

Effort.

In Farming.

In Life.

We must dig right in.

Listen.

Learn.

Laugh.

To see the cow pies.

And not just the shit.

LOL!

"Check!"

STORY 13:

We'll Get There

Discipline alone can set the pace for a long, fulfilling life.

Arm-in-arm in America's Dairyland.
Husband and wife.

Farmer and farm wife.

Who built a life.

Together.

Who celebrated simple.

Together.

An embodiment of the spirit of an industry and a people.

He in a cap and 'dairy' imprinted shirt.

Her beautiful white hair braided tightly, flowing down her shoulder.

Their arms.

Tightly crisscrossed.

Like their suspenders.

June Dairy Month.

Breakfast on the Farm.

Rounding the corner of the shed.

Slowly maneuvering the gravel drive.

People moving quickly by.

Same destination.

Through the yellow ribbons to the tent.

In line for a hearty breakfast.

Different pace.

Fast.

Slow.

One based on expediency.

One based on experience.

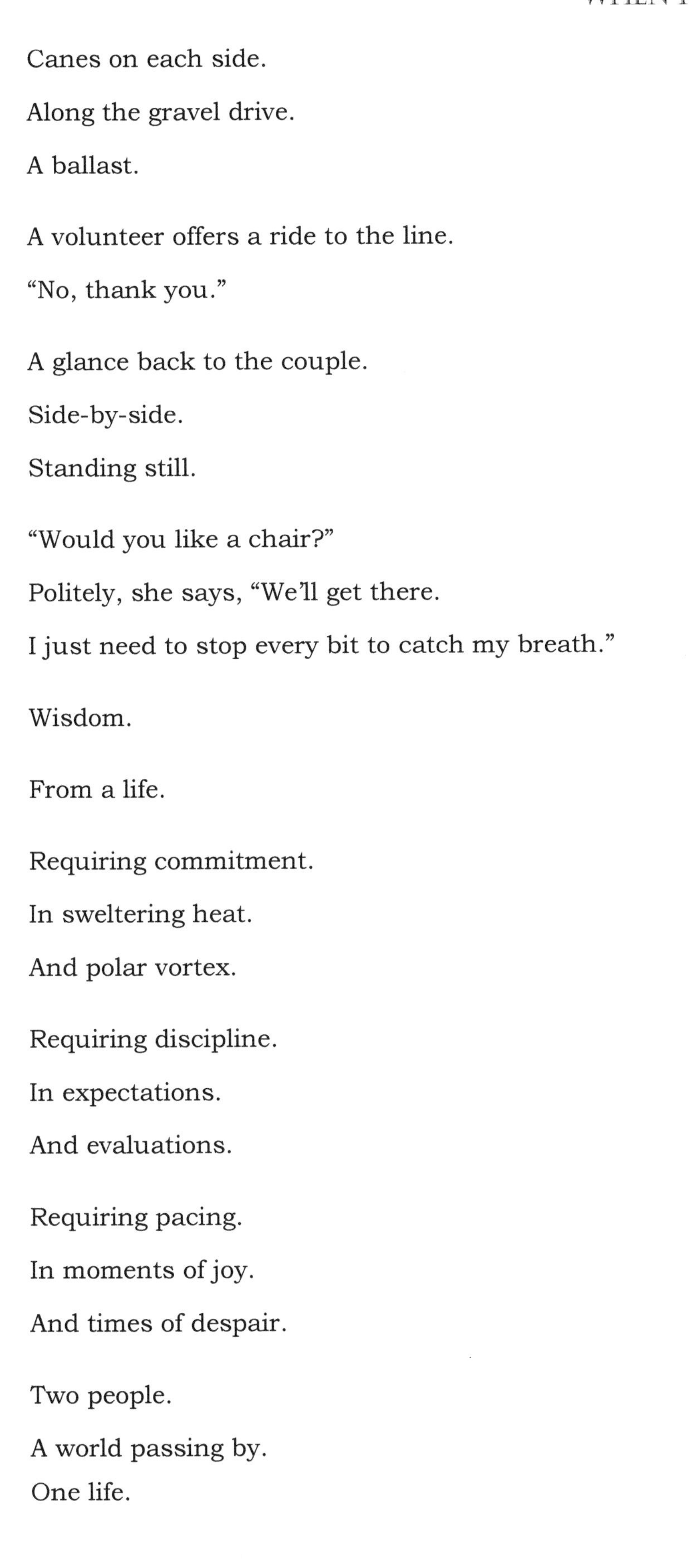

Canes on each side.

Along the gravel drive.

A ballast.

A volunteer offers a ride to the line.

"No, thank you."

A glance back to the couple.

Side-by-side.

Standing still.

"Would you like a chair?"

Politely, she says, "We'll get there.

I just need to stop every bit to catch my breath."

Wisdom.

From a life.

Requiring commitment.

In sweltering heat.

And polar vortex.

Requiring discipline.

In expectations.

And evaluations.

Requiring pacing.

In moments of joy.

And times of despair.

Two people.

A world passing by.

One life.

Co-workers.

Independent farmers.

Connected.

By purpose.

By pick-up schedule of the milk truck.

Perspective.

In a fast-paced world.

Interrupted.

Now.

Same.

Slow.

Pace.

Uncomfortable.

Wanting . . .

A return to 'simple'

Unaware.

Needing . . .

A return to 'simple.'

Understand.

Moving forward.

Catching breath.

All.

We.

Need.

Choose your pace.

Fulfilling . . .

Or futile . . .

STORY 14:

Born to Serve: The Farmer & the Marine

"How'd you get in here?'

And so, began a friendship:

the farmer and the Marine.

Listening:

Brian, a farm kid, summers filled with:

Pulling nails.

Straightening nails.

Re-using nails.

To build sheds.

A boy.

Bowl-haircut.

Bucket, at the ready.

We quickly learn.

Our parents had kids to handle chores around the house.

A farm kid learns his or her role!

Rock pickers.

Rock picking. A daily chore.

Brian and siblings create a competition.

"I can lift a half a bucket!"

For bragging rights.

Because "I'm all about the challenge!"

A life-long prize.

Discovery.

Vs.

Defeat.

Determination.

Vs.

Dread.

Destinations.

Vs.

Dead ends.

I knocked loudly on the shed door.

Walked in.

Replying: 'the gate was open.'

The gate works as intended to keep others out.

Brian knows what he wants.

Researching business ideas and questions.

Then, decisions.

Brian's dad is a farmer.

With an off-farm job as a custodian in a local school.

And he's always gathered scrap metal.

Three income sources.

Brian was provided for . . .

Choose to eat cow tongue for dinner or choose not to eat.

Brian ate.

When he asked for a new bike, his dad pointed to the scrap heap.

Brian made a new bike.

When Brian wanted his first truck, he already knew what to do.

He searched the line fence, put in $100 to make it run.

Brian saw that nothing in life comes to you.

Work is play.

A disciplined approach to life.

Instilled by his parents.

"A willingness to survive," he states.

After high school.

The Marines.

A decision to define who he is.

Through

Demonstration.

Distinction.

Self-determination.

Brian chose to become a logistics specialist.

"Beans. Bullets. And Band-Aids."

Essential elements of a Marine's sustenance.

His mantra for 'life.'

He knows what's essential to survive.

"I could out-wrestle everyone –

'cuz I had the will to live

and stamina to take the pain."

Listening . . .

I saw the clearest demonstration of why we do chores.

We are born to serve.

Family.

Country.

Self.

Nothing can be expected.

Everything must be earned.

U.S. Marine,

Brain became known as the guy who earned his rank responsibilities.

Most nights he hit the weight room.

To build strength.

For mind.

For body.

Thought: Are we duty-bound to serve ourselves?

Brian recalls, "I knew growing up I needed to work."

For individual sustainability.

For collective purpose.

Working together.

To take the hill.

To survive.

As a band of brothers.

“Improvise. Adapt. Overcome.”

The rallying cry of a Marine.

Quite a pivot for an ‘Insta-generation.’

Ponder with friends around a bonfire pit

and a few beers . . .

Are we ultimately then, beyond God,

responsible to be dependent on no one.

In the Marines,

I learned I could stand on my own two feet!

No ‘Gap Year’ for the farmer to ponder his existence.

School-to-farm.

Straight to work.

So, we may enjoy ‘Farm-to-Table.’

A farmer knows.

Survival is his mission.

Standard Operating Procedure.

For a Marine.

For a human.

Life.

Not. In fear.

Life in reality.

Fact Checked.

Confirmed.

The farmer knows.

Served.

Earned.

Sargent.

Citizen.

Leaning on a little something called humility and dues.

Skills.

Learned.

Diesel Tech.

Mechanic.

Concrete Layer.

Equipment painter.

Truck driver.

Farmer.

Income sources.

Diversified.

A back up plan.

Prepared to fight.

We don't have to rely on someone else.

A man.

A Marine.

A mission.

"We need to better ourselves."

For each other.

And so, continues the friendship with the farmer and the marine . . .

then he convinced me to pick rocks with him on his farm.

"Challenge!" Accepted.

STORY 15:

Markers that Bring You Home

If we just slow down to remember what comforts us,

we'll find home.

Landmarks.

Red boards.

And fieldstone foundations.

Outposts of Industry.

Along highways.

To let us know . . . we are almost home.

Named for 'founders' and families.

With meaning.

Settlers.

The Schneider Farm.

Now named for 'convenience and consumption.'

Replaceable.

Corporations.

Shopping Centers.

Markers.

And Meaning.

To remind us how to make decisions.

With confidence.

To persevere on a personal goal.

Like a Century Farm.

For 100 + years . . .

Brick School Road . . . Irish Valley Road . . .

we're almost home.

A simple, white farmhouse.

A trio of big blue silos.

The barn with arched doors.

Comfort.

In old landmarks.

Farms.

That created the possibilities we enjoy today.

Brick facades of Main St. that keep us connected to family members since gone.

In old-fashioned markers.

Values.

How we do things . . .

. . . lest we forget how they lived– survived

Or take our lives for granted.

Guideposts on the landscape . . . meticulously maintained.

How the son of a farmer was taught to tend the farm.

The yard light.

Let's late night travelers know . . . you aren't alone . . .

We're here . . . we're just recharg-ing.

We've got work to do when we wake.

Reflections on the landscape.

Silos.

As Sentinels.

Standing watch over us.

Make us feel safe.

Because they are still there.

Weathered & withering.

Restored & resilient.

Still there.

The memory.

Of the people we saw beyond the barbed wire fences.

. . . we're almost home.

They were always home.

Tireless.

Tenders.

Of Tradition.

They were there
when we needed
them.

Torn down.

Thrown out.

Heartache.

It's a modern world
. . . after all!

Modern Markers.

Shiny.

New.

High tech.

"Liked."

Selfies.

Instantly.

Snaps.

Retweets.

Filters.

Fake.

Unhappy.

Uncomfortable.

Un-authentic!

Choas.

Calamity.

Sadness.

Instantly.

From coast to coast.

OMG!

I can't handle this.

Look around.

Find.

Your markers.

Many. Still grounded. Where
they've always been.

Farmers.

Unchanged.

Changing world.

Still creating . . . a life. from dirt.

Fewer markers . . .

Some still standing alone like a disgraced monument.

Where.

A tree grows inside.

A stone foundation.

Windmills. Stuck.
Like a smile with
missing teeth.

Barns with sagging
roofs. Back-broken.

For Sale signs.

Generations re-
moved from a farm.

Wondering what's
coming next.

Unaware.

Knowledge.

Directions.

'How To Manuals.'

Of Work. Play. Living.

Markers on the landscape.

Remembrances for many.

Empty placeholders for the rest.

Now.

With certainty.

Rest. With your comfort.

Awaken to the work.

The opportunity.

To help your family find their markers.

You know. Meaning.

Work for possibilities. Or for paycheck.

Does work bring meaning?

What do the markers in your life say?

Mere structures on a landscape.

Life-sized emojis.

To structure our thinking.

Our decision-making.

For ourselves.

For our communities.

Meaningful.

Lane markers.

Some still standing.

Others, deposited in our memory banks.

Markers of living.

Discern those that bring your family solace.

. . . if we just slow down to remember what comforts us,

we'll find home.

PHOTO GLOSSARY

When I Listened to a Farmer

Cover:	Nick Raaum
Introduction:	Scott Mericka
Story 1:	Jim and Joyce Lins, Jessi Twardokus, Lori Lins, Jeff Lins.
Story 3:	Jennifer Brattlie, Brent Brattlie, Dennis Brattlie
Story 7:	Fran Rademacher, Alan Rademacher, Stuart Meier
Story 8:	Justin Krebs, Bruce Krebs
Story 9:	Jenna Langer, Jenifer Zimmerman, Brenda Von Rueden, Jessi & Nathan Twardokus, Kay MacLeish, Kelly Placke, Levi Wilke, Steve Soldner, David McCarthy, Quinton McCarthy, Steven Lins, Liana & Scott Mericka and children, Steven Taylor, Jarous Volenec, Charles Volenec, Nick Raaum, Chuck Uebersetzig
Story 10:	Pete Curran, Steve Soldner
Story 11:	Haily Soldner, Steve Soldner.
Story 12:	Pete Curran, (two curious cows)
Story 13:	Bill and JoAnn Rupnow
Story 14:	Brian Von Rueden

ACKNOWLEDGMENTS

Steven Lins . . . and family. From my first sales call on your farm, sharing pie and ice cream in your kitchen to ask questions about 'farming,' butchering chickens with your mom, feeding a newborn calf its first milk, milking cows in your parlor, installing the Stars & Stripes atop your silo . . . together, and of course collecting a fecal sample without gloves . . . I'm grateful you welcomed me – like most farmers do – onto your farm. Where I learned about your business. And your life.

ABOUT THE AUTHOR

Pete Curran has a big smile. Always did. As a kid, Pete was introspective, yet hyperactively-outgoing, and now is an unlikely 'author.' Dyslexic, Pete became a skimmer and avoided reading. He absorbs information and communicates visually.

Neither poetry, narrative or short story, Curran's stories sometimes find patterns – described as soliloquies, his visual recollections are delivered like he reads – staccato . . . one word or phrase at a time. Slow enough for him to see the words to understand them.

His storytelling takes a look at hard lessons from rural America and how they serve as solutions for younger generations in times of chaos.

As a city kid and 'outsider' to farming, Curran now embraces agriculture for the next generation. In 2019, he created the first-ever Interactive Ag Career Scavenger Hunt during Wisconsin Farm Technology Days—one of the largest public events in the Midwest.

The Rountree Gallery highlighted Curran's work at its 40th anniversary event in 2020. There, he shared *'a life. from dirt.'* on the struggles and successes of Wisconsin's farmers. His storytelling began in Junior High when he got his first SLR camera. Now with his iPhone and notebook, his stories grew from Facebook, When I Listened to a Farmer.

Having once rejected his father's blue-collar roots and dirty hands, Pete now celebrates farmers while working in agriculture and sharing what he hears and sees. In America's Dairyland. Where he wonders the value of physical work . . . on body and soul.

pete@ListenToAFarmer.com

CPSIA information can be obtained
at www.ICGtesting.com
Printed in the USA
BVHW020952171122
652193BV00017B/721